Self-Driving

Dreams

How AI Is Changing the Way We Drive

Etienne Psaila

Self-Driving Dreams: How AI Is Changing the Way We Drive

First Edition: **December 2024**

ISBN: 978-1-923393-18-9

Table of Contents

Chapter 1: Introduction: The Road Ahead

Imagine standing on the edge of a new world, where the hum of an engine is no longer accompanied by the familiar grip of a steering wheel, and the rhythm of traffic obeys not human whims but the precise logic of algorithms. This is the promise of autonomous driving: a reimagining of transportation that is as much about innovation as it is about transformation. For decades, the idea of self-driving cars has lived in the realm of science fiction, inspiring dreams of sleek machines whisking us to destinations while we read, work, or simply enjoy the view. Today, those dreams are beginning to merge with reality, and the journey toward full autonomy is already well underway.

Overview of Autonomous Driving Technology

At the heart of autonomous vehicles lies a sophisticated interplay of artificial intelligence, sensors, and vast amounts of data. These cars don't just drive; they "see," "think," and "learn." Using advanced technologies like LIDAR (Light Detection and Ranging), radar, cameras, and ultrasonic sensors, autonomous vehicles create a detailed, 360-degree view of their surroundings. AI processes this flood of information in real-time, allowing the vehicle to navigate complex environments, avoid obstacles, and make split-second decisions.

But the journey to autonomy isn't solely about technology. It's a story of collaboration between software engineers, automotive giants, tech startups, and regulators, all working together to build a new era of transportation. It's a story where silicon chips become as important as steel frames and where code replaces combustion as the driving force.

Why Now? The Convergence of AI, Transportation, and Societal Need

Why is the world suddenly abuzz with talk of autonomous vehicles? The answer lies at the intersection of technological breakthroughs and pressing societal challenges.

Advances in AI and machine learning have enabled machines to perform tasks once thought impossible, from recognizing human emotions to mastering complex games like Go. In the automotive world, these same technologies are being harnessed to teach vehicles how to drive. At the same time, the proliferation of high-speed internet and 5G networks has created an ecosystem where cars can communicate with each other and with smart infrastructure.

Yet the push for self-driving cars is about more than just cutting-edge tech. Society is grappling with urgent problems that autonomous vehicles promise to solve. Road safety is a prime example. According to the World

Health Organization, over 1.3 million people die each year in road traffic crashes, with human error responsible for 94% of those accidents. Autonomous vehicles, with their unblinking focus and instantaneous reaction times, offer the potential to drastically reduce this toll.

Environmental concerns also play a role. With the rise of electric and autonomous vehicles, we're not just talking about a transportation revolution but a chance to rethink urban planning, reduce carbon emissions, and create greener cities.

The economic implications are equally profound. From ridesharing to logistics, autonomous driving could redefine industries, create new markets, and disrupt old ones. It's not just cars we're reinventing; it's the entire fabric of how we move, work, and live.

Defining the Autonomous Revolution

The shift to autonomy is not just incremental—it's revolutionary. Unlike other technological leaps, it touches every aspect of our daily lives. It challenges our notions of control, responsibility, and trust. Who is to blame if an autonomous car makes a fatal mistake? How do we legislate an algorithm? And perhaps most provocatively, what happens to our sense of freedom when the act of driving—a quintessential human experience—is handed over to machines?

This revolution is not unfolding in isolation. It is being shaped by a broader wave of technological innovation: the rise of artificial intelligence, the spread of connectivity through the Internet of Things, and the relentless pursuit of sustainability in the face of climate change. Together, these forces are creating a perfect storm, one that is propelling us toward a future where mobility is safer, smarter, and more efficient.

But revolutions are rarely smooth. There will be roadblocks—technical, ethical, and societal. There will be debates about the trade-offs we're willing to make and the pace of change we can tolerate. This book will delve into these complexities, offering a roadmap to understand where we are, how we got here, and where we might be headed.

As we embark on this journey, one thing is clear: the road ahead is unlike any we have traveled before. Autonomous vehicles are not just reshaping transportation; they are redefining what it means to move through the world. And as with any great journey, the destination is only part of the story. The path we take to get there will define the legacy of this moment in history.

Let's begin.

Chapter 2: A Brief History of Autonomy

The idea of cars driving themselves has captured human imagination for over a century, long before the technology to achieve it existed. From the pages of science fiction to the labs of innovative engineers, the journey toward autonomy is as much about dreaming big as it is about solving complex problems. Understanding this history is key to appreciating the profound changes on the horizon.

The Early Dreams of Self-Driving Cars: Science Fiction to Prototypes

In the early 20th century, as automobiles became symbols of progress, the dream of self-driving cars began to emerge. It was a world of boundless optimism, where technology was believed to be the key to solving humanity's greatest challenges. Writers and futurists envisioned cities bustling with vehicles that required no human input, freeing us from the drudgery of driving.

One of the first depictions of autonomous vehicles appeared in the 1939 World's Fair in New York. General Motors presented "Futurama," an exhibit that featured a vision of a world dominated by highways and automated cars. At the time, it seemed more like a utopian fantasy than a realistic projection. Yet, it planted a seed.

The 1950s and 60s saw experiments that brought these fantasies closer to reality. In 1956, General Motors showcased the Firebird II, a concept car designed to operate on an automated highway system. Meanwhile, engineers began experimenting with sensors and rudimentary guidance systems. These early attempts were crude by today's standards, but they demonstrated that autonomous driving wasn't just a dream—it could be a pursuit.

Key Milestones in the Development of Autonomous Vehicle Technology

The road to autonomy was paved with technological milestones, each pushing the boundaries of what was possible. In the 1980s, a major breakthrough came from Carnegie Mellon University's NavLab project. Their self-driving vehicle used cameras and computers to process road images and navigate. Around the same time, Ernst Dickmanns and his team at the University of Bundeswehr Munich developed the VaMoRs, a Mercedes van outfitted with cameras and sensors capable of lane-keeping and obstacle detection.

The 1990s saw further refinement. Researchers began to incorporate GPS, which provided vehicles with a reliable way to determine their position. Companies like Toyota and Honda also began exploring driver-assistance systems, planting the seeds for technologies like adaptive cruise control and lane-keeping assistance.

But it wasn't until the 2000s that the quest for autonomy entered the mainstream. The rise of artificial intelligence and machine learning allowed cars to interpret vast amounts of data in real time. Technologies like LIDAR and advanced computer vision systems gave vehicles a way to "see" their environment with unprecedented accuracy. The dream of a fully autonomous car was no longer a question of "if," but "when."

The DARPA Challenge and Its Significance

A pivotal moment in the history of self-driving cars came in 2004 with the first DARPA (Defense Advanced Research Projects Agency) Grand Challenge. This competition was a bold experiment: teams from around the world were invited to design vehicles capable of navigating a 150-mile course through the Mojave Desert—without any human intervention. The results were both humbling and inspiring. None of the vehicles completed the course, with the best performing car managing only seven miles before breaking down.

Despite the failure, the challenge ignited a wave of innovation. It brought together engineers, computer scientists, and roboticists, many of whom would go on to become key figures in the development of autonomous vehicles. DARPA's subsequent challenges in 2005 and 2007 were even more successful. In 2005, five vehicles completed the course, and in 2007, the Urban Challenge demonstrated self-driving cars navigating simulated city environments, obeying traffic laws, and avoiding collisions.

These competitions showcased the power of collaboration and open competition. They proved that the technology was no longer a distant dream but a tangible reality. The breakthroughs achieved during the DARPA challenges laid the foundation for the autonomous vehicle programs we see today.

A Legacy of Innovation

The history of autonomy is a testament to human ingenuity and perseverance. From the speculative musings of science fiction writers to the groundbreaking experiments of DARPA, each chapter in this story has brought us closer to a world where cars drive themselves. What once seemed like a whimsical dream is now a reality being shaped by engineers, designers, and visionaries.

This brief history sets the stage for the rest of our journey. The next chapters will delve into the nuts and bolts of how these vehicles work, the challenges they face, and the opportunities they present. The story of autonomy is still unfolding, and the best is yet to come.

Chapter 3: The Pioneers of Self-Driving Technology

The road to autonomous vehicles has been paved by visionaries, institutions, and companies daring enough to transform bold ideas into groundbreaking realities. From tech giants to scrappy startups, the pioneers of self-driving technology have propelled us into an era of innovation, competition, and disruption. Their stories are a testament to ambition, ingenuity, and the relentless pursuit of progress.

Profiles of Key Players

Google (Waymo): The Visionary Leader

In 2009, Google's self-driving car project, later spun off as Waymo, became the vanguard of autonomous vehicle development. Spearheaded by Sebastian Thrun, a Stanford professor and co-creator of Google Street View, Waymo envisioned a world where transportation was safer and more accessible. Using AI, LIDAR, and high-definition mapping, Waymo vehicles began logging millions of miles on public roads. Their focus wasn't just on technology but also on perfecting real-world performance, making them the industry benchmark. Today, Waymo operates one of the most advanced fleets of fully autonomous vehicles, leading the charge toward a driverless future.

Tesla: Redefining the Driving Experience

Under Elon Musk's leadership, Tesla has become synonymous with innovation in electric and autonomous vehicles. While Tesla's approach to autonomy differs from Waymo's—favoring camera-based systems and neural networks over LIDAR—it has been no less influential. The introduction of Tesla's Autopilot in 2015 gave consumers a glimpse of an autonomous driving experience, though still requiring driver oversight. Tesla's bold promises, like achieving full self-driving capabilities, have both inspired confidence and sparked controversy, positioning the company as both a disruptor and a lightning rod in the autonomous space.

Uber: Pioneering Autonomy in Ridesharing

For Uber, self-driving technology represented not just innovation but survival. Faced with the rising costs of human drivers, the ridesharing giant invested heavily in autonomy through its Advanced Technologies Group (ATG). While Uber's journey was marred by challenges—including a fatal accident involving an autonomous test vehicle in 2018—the company's research and development pushed the boundaries of urban autonomy. Though Uber sold ATG to Aurora Innovation in 2020, its early experiments highlighted the complexities and potential of deploying autonomous vehicles in dense, real-world environments.

Other Notable Players

- **GM Cruise:** General Motors' subsidiary Cruise has focused on building all-electric, autonomous vehicles designed specifically for urban environments.

- **Apple:** Though secretive about its progress, Apple's "Project Titan" is widely rumored to be exploring autonomy, leveraging its expertise in software and design.

- **Baidu:** China's tech giant Baidu is advancing autonomous technology in Asia, with ambitious plans for commercial robotaxis.

Individual Visionaries

Elon Musk: The Maverick Innovator

Few names are as polarizing and influential as Elon Musk. Tesla's CEO has not only pushed the boundaries of electric vehicle technology but has also positioned autonomy as a cornerstone of his vision. Musk's focus on a scalable, camera-driven approach has sparked debates within the industry, but his bold claims—such as enabling Tesla cars to operate as autonomous ridesharing vehicles—have kept the public and competitors on edge. Love him or hate him, Musk's relentless ambition has undeniably accelerated the autonomous race.

Chris Urmson: The Steady Trailblazer

As a lead engineer for Google's self-driving car project and later co-founder of Aurora Innovation, Chris Urmson has been a quiet yet impactful figure in autonomy. Known for his methodical approach, Urmson's work has focused on creating safe and scalable autonomous solutions. His vision balances technological optimism with a realistic understanding of the challenges ahead.

Anthony Levandowski: The Controversial Genius

Few figures embody the tumultuous nature of the autonomous revolution like Anthony Levandowski. As a key player in both Google's Waymo and Uber's self-driving efforts, Levandowski contributed significantly to early advancements. However, his career was overshadowed by legal battles, including accusations of stealing trade secrets. Despite his controversies, Levandowski's contributions remain foundational to the industry's development.

The Role of Startups and Academic Institutions

While tech giants dominate headlines, startups and universities have played an equally critical role in shaping autonomous technology.

Startups: The Agile Innovators

Startups like Zoox, Aurora, and Nuro have injected fresh

ideas into the industry. Zoox, for instance, is reimagining vehicles entirely, designing bidirectional, electric robotaxis optimized for urban use. Meanwhile, Nuro focuses on delivery robots, demonstrating that autonomy has applications far beyond passenger transport. These companies, unencumbered by legacy systems, have the flexibility to explore niche markets and novel approaches.

Academic Institutions: The Birthplace of Innovation

From Carnegie Mellon University to Stanford, academic research has laid the groundwork for many of the technologies powering autonomous vehicles today. The DARPA Grand Challenges, often spearheaded by university teams, catalyzed much of the industry's early progress. These institutions continue to train the next generation of engineers, ensuring that the autonomous revolution has a robust pipeline of talent.

A Landscape of Collaboration and Competition

The pioneers of self-driving technology are not just competitors; they are collaborators in a global effort to redefine mobility. Their contributions have turned a futuristic concept into a tangible reality, though the journey is far from over. As the industry evolves, the interplay between established players, startups, and academia will shape the future of transportation—and ultimately, how we live.

Chapter 4: How Autonomous Vehicles Work

The magic of self-driving cars lies in their ability to see, think, and act—much like a human driver, but with the precision and speed of advanced technology. This chapter delves into the inner workings of autonomous vehicles, revealing the fascinating interplay of sensors, artificial intelligence, and data that powers them.

The AI Behind the Wheel: Sensors, LIDAR, and Computer Vision

Imagine driving blindfolded—an impossible feat without the ability to perceive your surroundings. Autonomous vehicles face a similar challenge, relying on a suite of sensors to "see" the world. These sensors provide the raw data that artificial intelligence processes to navigate safely and effectively.

- **LIDAR (Light Detection and Ranging):** Often called the "eyes" of autonomous vehicles, LIDAR uses laser beams to create a detailed 3D map of the environment. By emitting pulses of light and measuring their return time, LIDAR builds an accurate representation of obstacles, distances, and terrain. It's especially effective at detecting objects in complex settings, like crowded intersections or dark conditions.

- **Cameras:**
  While LIDAR provides depth perception, cameras capture visual information—like road signs, lane markings, and traffic signals. Advanced computer vision algorithms enable the vehicle to interpret this data, identifying objects like pedestrians, cyclists, and other vehicles.

- **Radar and Ultrasonic Sensors:**
  Radar excels at detecting objects in motion, particularly in poor visibility conditions like fog or heavy rain. Ultrasonic sensors, often used for parking and close-range detection, add another layer of safety by identifying nearby obstacles.

These sensors work together to provide a comprehensive view of the vehicle's surroundings. However, collecting data is only half the battle. The real challenge lies in interpreting and acting on this information.

Mapping and Data-Driven Decision-Making

For an autonomous vehicle, the journey begins long before the engine starts. High-definition maps are a critical component, serving as a detailed blueprint of the world. These maps go far beyond traditional GPS; they include information about road gradients, curvature, lane configurations, and even the location of stop signs.

- **Building the Map:**
  Companies like Waymo and Tesla continuously update their maps using data collected from test vehicles and user fleets. This data is combined with satellite imagery and infrastructure databases to create an ultra-precise representation of the driving environment.

- **Localization:**
  Autonomous vehicles use their sensors to compare real-time observations with preloaded maps. This process, called localization, allows the car to determine its exact position on the road, down to a few centimeters.

- **Dynamic Decision-Making:**
  Beyond maps, self-driving cars rely on real-time data to make decisions. For instance, an unexpected construction zone or a pedestrian jaywalking requires the vehicle to adapt instantaneously. AI systems process this information using complex algorithms, weighing variables like speed, distance, and potential risks to choose the safest course of action.

Real-World Testing and Machine Learning Algorithms

The heart of autonomy is artificial intelligence, and its development hinges on one critical ingredient: data.

Autonomous vehicles are trained using machine learning algorithms, which improve over time as they encounter new scenarios.

- **Simulated Learning Environments:** Before testing on public roads, developers use simulations to expose autonomous vehicles to a wide range of driving scenarios. These virtual environments allow the AI to learn how to respond to rare but critical situations, such as a child chasing a ball into the street or a sudden vehicle breakdown.

- **On-Road Testing:** Real-world testing is where theory meets practice. Companies deploy fleets of test vehicles equipped with extensive recording equipment, gathering millions of miles of driving data. These vehicles encounter diverse conditions—rural roads, bustling cities, and extreme weather—to ensure reliability across all environments.

- **Continuous Improvement:** Machine learning enables vehicles to learn from their experiences. When an autonomous car encounters a new type of obstacle, the AI processes the event and integrates the solution into its decision-making framework. This knowledge is then shared across the entire fleet, creating a collective learning system.

The Balancing Act: Precision Meets Complexity

Building a self-driving car is a balancing act between precision and complexity. Sensors must deliver accurate data, but too much information can overwhelm the system. AI must process this data quickly enough to react to sudden changes, but it must also ensure every decision is safe. The interplay between these elements is what makes autonomous vehicles both a marvel of engineering and a formidable technical challenge.

The journey of how autonomous vehicles work reveals not only the ingenuity of the technology but also the intricacy of the problems it solves. It's a testament to human innovation, demonstrating how far we've come in our quest to automate one of the most fundamental aspects of modern life: driving.

Chapter 5: From Manual to Machine: The Levels of Autonomy

The journey from manually operated vehicles to fully autonomous cars isn't an overnight leap but a carefully mapped progression. The Society of Automotive Engineers (SAE) has defined five levels of vehicle autonomy, providing a clear framework to understand where we are today and what lies ahead. Each level represents a milestone in the gradual evolution from human-driven cars to machines that navigate the world without any human input.

Explaining SAE's Five Levels of Vehicle Autonomy

Level 0: No Automation

This is the starting point—vehicles with no autonomous features. At this level, the driver is in full control, performing all tasks like steering, braking, and accelerating. Think of classic cars or even modern vehicles without any advanced driver-assistance systems (ADAS).

Level 1: Driver Assistance

At Level 1, automation begins to make its presence felt. These vehicles include systems that assist the driver with specific tasks, such as adaptive cruise control or lane-keeping assistance. The driver remains fully engaged and responsible for the vehicle but benefits from limited support to reduce workload.

- **Example:** A car that maintains a set speed and adjusts for traffic but requires the driver to steer.

Level 2: Partial Automation

Level 2 introduces more advanced capabilities, allowing the vehicle to control multiple functions simultaneously, such as steering and acceleration. However, the driver must remain attentive and ready to take over at any moment. Tesla's Autopilot and General Motors' Super Cruise are examples of Level 2 systems.

- **Key Feature:** The car can perform certain tasks but still relies heavily on human oversight.

Level 3: Conditional Automation

At Level 3, the vehicle can handle all driving tasks under specific conditions, such as highway driving. The driver can disengage from controlling the car but must be ready to intervene if the system encounters a situation it cannot handle. This level marks a significant step toward full autonomy but introduces legal and ethical challenges regarding liability and safety.

- **Current Example:** Honda's Legend sedan, approved for limited Level 3 driving in Japan.

Level 4: High Automation

Level 4 vehicles are capable of full autonomy within defined parameters, such as geofenced areas or specific routes. Unlike Level 3, the system does not require human intervention in these scenarios. However, outside its operational design domain (e.g., highways or urban centers), the vehicle may require manual operation or refuse to drive.

- **Key Feature:** The car is autonomous in certain conditions but not universally.

Level 5: Full Automation

At Level 5, the dream of autonomy is fully realized. These vehicles are designed to operate without any human input, in all environments and under all conditions. They have no need for steering wheels or pedals, as they are entirely self-reliant.

- **The Holy Grail:** Level 5 represents a world where every car on the road could theoretically drive itself, enabling transformative changes in urban planning, logistics, and human mobility.

Where the Industry Is Today

As of now, most vehicles on the market fall between Level 1 and Level 2. Automakers like Tesla and Mercedes-Benz have pushed the envelope with Level 2+

systems that flirt with conditional automation. However, true Level 3 and Level 4 autonomy remains in its infancy, with only a few experimental deployments, primarily in controlled environments like robotaxi services.

The leap to Level 5 autonomy is still years—if not decades—away. Challenges such as regulatory approval, technological limitations, and societal acceptance mean that full automation remains more of a vision than an immediate reality.

Transition Challenges Between Levels

The journey through these levels is fraught with challenges, as each step introduces new complexities.

- **Technological Hurdles:** Moving from Level 2 to Level 3 involves a seismic shift in responsibility, requiring systems that can accurately detect when human intervention is needed and reliably alert the driver. The technology must be fail-proof, as even minor errors can lead to catastrophic consequences.

- **Regulatory and Legal Barriers:** Level 3 autonomy raises critical questions about liability. If an accident occurs while the car is in control, who is responsible—the driver, the automaker, or the software provider? Governments around the world are grappling with these issues, slowing the path to widespread adoption.

- **Human   Factors:** One of the most overlooked challenges is human behavior. Transitioning between levels of autonomy often leads to over-reliance on systems or driver disengagement. The 2018 fatal crash involving a Tesla driver who was distracted while using Autopilot highlights the dangers of misunderstanding the system's limitations.

- **Economic   and   Social   Impacts:** Moving toward higher levels of autonomy also threatens to disrupt industries like trucking, taxi services, and auto manufacturing. While the benefits are clear, the transition requires addressing the economic dislocation it could cause.

The Road Ahead

The levels of autonomy represent more than just a technical roadmap—they symbolize humanity's growing trust in machines. While Level 5 autonomy is the ultimate goal, each intermediate level brings its own set of innovations and challenges. The industry's progress is a testament to our ambition, but it is also a reminder of the complexities involved in relinquishing control of something as fundamental as driving.

As we move further into this transition, the question becomes not just about whether technology can achieve full autonomy, but whether society is ready for the profound changes it will bring.

Chapter 6: Ethical Dilemmas on the Road

Autonomous vehicles promise to revolutionize transportation by reducing accidents and improving efficiency, but they also bring a Pandora's box of ethical challenges. These dilemmas force us to confront uncomfortable questions about trust, responsibility, and the moral framework of machines. As we shift control from human drivers to artificial intelligence, we must grapple with how these systems make decisions in life-or-death situations, who bears responsibility when things go wrong, and how biases in AI could perpetuate inequalities on the road.

The "Trolley Problem" in Autonomous Driving

The "trolley problem" is a classic ethical thought experiment that asks: if a runaway trolley is heading toward five people tied to a track, but you could divert it to a track with only one person, what should you do? It's a dilemma with no perfect answer—any choice results in harm. In the context of autonomous driving, the trolley problem becomes chillingly real.

Imagine a self-driving car faced with an unavoidable accident: swerve to avoid hitting a pedestrian but risk the lives of its passengers, or prioritize the safety of those inside the vehicle at the expense of bystanders. How should the AI decide? Unlike human drivers, whose split-second reactions are guided by instinct, autonomous systems must follow pre-programmed algorithms, forcing automakers and programmers to make these moral choices in advance.

- **Cultural Variations:** A study from the MIT Moral Machine project revealed that ethical preferences for such scenarios vary across cultures. For example, some societies prioritize the young over the elderly, while others emphasize saving as many lives as possible, regardless of age. These cultural differences complicate the design of universal ethical frameworks for self-driving cars.

- **Legal and Commercial Considerations:** Automakers may face conflicting incentives. Should they prioritize passenger safety, since those passengers are their paying customers? Or should they prioritize the greater good to minimize societal harm? Striking a balance between morality, legality, and marketability is an ongoing challenge.

Responsibility in Accidents: Human vs. Machine

One of the most contentious questions surrounding autonomous vehicles is responsibility. When a human driver causes an accident, fault is often clear. But when an autonomous vehicle is involved, determining liability becomes a labyrinth of ethical, legal, and technological questions.

- **The Manufacturer's Role:** If an autonomous vehicle's decision-making algorithm causes an accident, should the manufacturer bear responsibility? This question shifts the legal landscape from individual liability to corporate accountability, with automakers potentially facing lawsuits for system failures.

- **The Human Element:** In Level 2 and Level 3 systems, where drivers are still expected to intervene when necessary, ambiguity arises. If the car alerts the driver to take control but they fail to respond in time, is the fault theirs, or does the system bear some responsibility for not predicting the situation sooner?

- **Insurance Implications:** Autonomous vehicles are already prompting shifts in the insurance industry. Policies may need to evolve to account for machine errors, cyberattacks, or even miscommunication between different autonomous systems on the road.

- **Precedents and Case Studies:** Real-world incidents have already highlighted these challenges. In 2018, a self-driving Uber struck and killed a pedestrian in Arizona. Investigations revealed both human oversight failures and deficiencies in the car's programming. The case underscored the urgent need for clearer guidelines on liability and responsibility.

Bias in AI Systems and Implications for Road Safety

Artificial intelligence is only as unbiased as the data it learns from—and in many cases, that data reflects societal inequalities. If not carefully designed, the AI behind autonomous vehicles could inadvertently perpetuate or even exacerbate these biases.

- **Unequal Safety Outcomes:** Studies have shown that facial recognition systems often perform worse when identifying individuals with darker skin tones. If similar biases exist in AI systems for pedestrian detection, certain groups could face greater risks in interactions with autonomous vehicles.

- **Infrastructure   Disparities:**
Autonomous systems rely on well-maintained
roads, clear lane markings, and consistent
signage—conditions often lacking in lower-
income neighborhoods. This disparity could
mean that self-driving cars perform less reliably
in underserved areas, reinforcing existing
inequalities.

- **Economic   Access:**
If autonomous vehicles remain expensive or
require subscription-based services, they may be
inaccessible to lower-income communities,
further widening the mobility gap.

- **Regulating   Bias:**
To combat these issues, developers must
rigorously audit their AI systems for bias and
ensure that training data represents a diverse
range of environments and scenarios.
Governments and watchdog organizations also
have a role to play in setting standards and
enforcing accountability.

The Moral Crossroads

Autonomous vehicles challenge us to rethink what it means to trust machines with ethical decisions. Unlike humans, who navigate moral dilemmas with a mix of emotion, intuition, and societal norms, AI operates within the rigid boundaries of its programming. As such, the ethical dilemmas of self-driving cars are not just technical problems—they are reflections of our values as a society.

Resolving these issues requires collaboration between engineers, ethicists, policymakers, and the public. How we choose to address them will shape not only the future of autonomous vehicles but also the broader relationship between humanity and artificial intelligence.

Chapter 7: Regulating the Revolution

The rise of autonomous vehicles represents one of the most profound shifts in modern transportation, but with great innovation comes the need for equally transformative regulation. Governments around the world are grappling with the challenge of fostering the growth of self-driving technology while ensuring public safety, addressing liability concerns, and protecting privacy. The regulatory framework for autonomous vehicles isn't just a set of rules—it's a balancing act that could determine how quickly, safely, and equitably this technology is adopted.

Global Approaches to Self-Driving Legislation

Different regions have adopted diverse strategies for regulating autonomous vehicles, reflecting varying levels of technological readiness, infrastructure, and societal attitudes.

United States: A Patchwork of Policies

In the U.S., regulation of autonomous vehicles is largely fragmented. While the federal government, through agencies like the National Highway Traffic Safety Administration (NHTSA), sets broad guidelines, individual states have significant authority over vehicle laws. States like California, Arizona, and Nevada have become testing hubs due to their progressive policies, but this patchwork approach creates inconsistencies.

- **Strengths:** Flexibility allows states to experiment with different approaches to regulation.

- **Weaknesses:** Lack of uniformity complicates nationwide deployment and creates confusion for automakers and developers.

European Union: A Unified Framework with Local Nuances

The EU has taken a more cohesive approach, emphasizing safety and data protection through regulations like the General Data Protection Regulation (GDPR). The EU's Automated Driving Roadmap provides clear guidelines for testing and deploying autonomous vehicles, but individual countries retain some autonomy, leading to variations in implementation.

- **Strengths:** Strong emphasis on ethical AI and privacy.

- **Weaknesses:** Slower innovation due to stringent regulations.

China: Rapid Advancement Under Centralized Oversight

China has embraced autonomous vehicles with a top-down regulatory approach. Backed by government support, companies like Baidu and Pony.ai have benefited from designated testing zones and accelerated approval processes. The government's focus on urban planning also integrates self-driving cars into smart city initiatives.

- **Strengths:** Centralized control allows for rapid progress and large-scale testing.

- **Weaknesses:** Concerns about data privacy and government surveillance.

Other Notable Approaches

- **Japan:** Prioritizes public safety and accessibility, particularly for elderly populations.

- **Australia:** Focuses on harmonizing national and state regulations while preparing infrastructure for autonomy.

Challenges with Insurance, Liability, and Data Privacy

Insurance: Rethinking Risk Models

Traditional insurance models are built on the assumption of human drivers, but autonomous vehicles challenge this paradigm.

- **Shifting Liability:** As control shifts from drivers to algorithms, insurers must determine who is liable—car manufacturers, software developers, or fleet operators.

- **New Risks:** Autonomous vehicles introduce unique risks, such as software bugs or cyberattacks, that require specialized insurance policies.

- **Opportunities:** Reduced accident rates could lower overall premiums, benefiting consumers.

Liability: The Blame Game

Determining fault in accidents involving autonomous vehicles is a complex legal puzzle.

- **Scenario 1:** If an autonomous car crashes due to a sensor failure, is the manufacturer responsible?

- **Scenario 2:** If a human driver fails to intervene when alerted, does the fault lie with the driver or the system?

- **Emerging Models:** Some propose strict liability for manufacturers, while others suggest shared responsibility based on the circumstances.

Data Privacy: Who Owns the Data?

Autonomous vehicles generate vast amounts of data, including location, driving behavior, and passenger information. This data is critical for improving AI systems but raises significant privacy concerns.

- **Ethical Questions:** Who owns the data collected by self-driving cars—the owner, the manufacturer, or a third-party operator?

- **GDPR and Beyond:** Europe's GDPR sets a high standard for data protection, but other regions have yet to implement comparable safeguards.

- **Transparency:** Consumers must be informed about how their data is used, stored, and shared.

The Role of Governments in Fostering Innovation While Ensuring Safety

Governments face the dual challenge of encouraging innovation while protecting public safety. Striking this balance requires a nuanced approach.

- **Encouraging Innovation:**

  - **Subsidies and Grants:** Providing financial support for R&D in autonomous technology.

  - **Testing Zones:** Creating designated areas for companies to test self-driving cars in real-world conditions.

  - **Public-Private Partnerships:** Collaborating with tech companies and automakers to advance infrastructure and standards.

- **Ensuring Safety:**

  - **Certification Processes:** Establishing rigorous testing and certification requirements before vehicles can be deployed.

  - **Ethical Guidelines:** Developing frameworks to address moral dilemmas and AI bias.

  - **Public Education:** Educating citizens about the capabilities and limitations of autonomous vehicles to build trust.

- **Global Collaboration:**

  - The cross-border nature of autonomous

technology demands international cooperation. Forums like the United Nations Economic Commission for Europe (UNECE) are already working on harmonizing standards, but greater collaboration is needed to address shared challenges.

Navigating the Road Ahead

Regulating autonomous vehicles is as much about shaping the future as it is about responding to the present. How governments navigate these challenges will determine not only the speed of adoption but also the societal impact of this technology. By creating a framework that fosters innovation, ensures safety, and protects privacy, policymakers have the opportunity to steer the autonomous revolution in a direction that benefits everyone.

Chapter 8: AI Meets Infrastructure

The promise of autonomous vehicles is deeply tied to the environments in which they operate. For self-driving cars to function seamlessly, they need more than just advanced onboard technology—they require smart infrastructure that supports their navigation, communication, and integration with existing systems. This chapter explores how AI-driven transportation demands smarter cities, the synergy between autonomous vehicles and public transit, and the critical role of technologies like 5G and the Internet of Things (IoT) in vehicle-to-everything (V2X) communication.

The Need for Smart Cities to Support Autonomous Vehicles

The infrastructure that underpins traditional driving is ill-equipped to handle the demands of autonomous systems. Smart cities, with their interconnected networks and data-driven systems, offer the solution. By integrating AI into urban planning and infrastructure, cities can create an environment where autonomous vehicles thrive.

- **Intelligent Traffic Management:** Smart traffic signals equipped with sensors and AI algorithms can adapt to real-time traffic conditions, reducing congestion and improving flow. These systems communicate with autonomous vehicles, providing critical information about upcoming traffic patterns or hazards.

- **Enhanced Road Markings and Signage:** Clearly marked lanes and digital signage that can update dynamically are essential for AVs to interpret their surroundings accurately. Smart infrastructure can relay this information directly to the vehicle's systems.

- **Autonomous-Only Lanes and Zones:** As autonomous vehicles become more common, cities may designate specific lanes or areas for their use, ensuring smoother operation and minimizing interaction with human drivers.

- **Data Ecosystems:** Smart cities generate vast amounts of data through sensors, cameras, and IoT devices. This data can be shared with autonomous vehicles to enhance their decision-making, from real-time weather conditions to pedestrian density.

The transformation into smart cities requires significant investment and coordination between governments, private companies, and urban planners. However, the benefits—safer roads, reduced congestion, and cleaner environments—justify the effort.

Integration with Public Transportation and Urban Planning

Autonomous vehicles should not replace public transportation but complement it. Integrating self-driving technology with existing transit systems can create a more efficient and equitable urban mobility network.

- **First-Mile/Last-Mile Solutions:** One of the biggest barriers to using public transit is the difficulty of reaching stations or stops. Autonomous shuttles or ride-hailing services can bridge this gap, making public transportation more accessible and attractive.

- **Dynamic Routing and Scheduling:** AI can optimize public transit routes and schedules based on real-time demand, ensuring that buses, trains, and autonomous shuttles operate at maximum efficiency.

- **Redesigning Urban Spaces:** The widespread adoption of autonomous vehicles could reduce the need for parking lots and wide roads, freeing up valuable land for parks, housing, and community spaces. Urban planners must consider these changes when designing the cities of the future.

- **Autonomous Buses and Trams:** Some cities are already exploring fully autonomous public transit. For example, autonomous buses could operate on fixed routes with high reliability, offering an eco-friendly alternative to private cars.

By aligning autonomous technology with public transportation, cities can reduce car dependency, lower emissions, and promote a more sustainable urban lifestyle.

The Role of 5G and IoT in Vehicle-to-Everything (V2X) Communication

At the heart of autonomous vehicles' ability to navigate and respond to their environment is communication. V2X technology enables vehicles to exchange information with other vehicles (V2V), infrastructure (V2I), and even pedestrians (V2P). This ecosystem relies heavily on 5G and IoT to deliver the speed, reliability, and connectivity required.

- **5G: The Backbone of V2X Communication**
  The high speed and low latency of 5G networks are essential for real-time communication between vehicles and their surroundings. For example:

    o A car approaching an intersection can receive data from traffic signals, pedestrian crossings, and nearby vehicles to determine the safest path forward.

- In emergencies, 5G enables instant alerts about accidents or road closures, allowing vehicles to reroute immediately.

- **IoT-Driven Smart Infrastructure:** IoT devices embedded in roads, signs, and traffic signals collect and share data with autonomous vehicles, creating an interconnected network. Examples include:

  - Sensors that detect road surface conditions, such as ice or potholes, and relay this information to approaching vehicles.

  - Smart parking systems that guide vehicles to available spaces, reducing time spent searching and lowering emissions.

- **Improving Safety and Efficiency:** V2X communication can significantly reduce accidents by enabling vehicles to "see" beyond their sensors. For example, a car can receive warnings about a vehicle braking suddenly two cars ahead, allowing it to react preemptively.

- **Enabling Coordination in Mixed Traffic:** In environments with both autonomous and human-driven vehicles, V2X communication ensures that AVs can predict and adapt to human behavior more effectively, minimizing risks.

The Challenges of Building Smart Infrastructure

While the benefits of smart cities and V2X communication are clear, there are significant barriers to implementation:

- **Cost:**
  Upgrading existing infrastructure and deploying 5G networks is expensive, especially for developing nations or cash-strapped municipalities.

- **Coordination:**
  Creating smart cities requires collaboration between multiple stakeholders, including governments, private companies, and local communities. Aligning their interests can be complex.

- **Cybersecurity   Risks:** As vehicles and infrastructure become more connected, they become targets for cyberattacks. Robust security measures are critical to prevent hacking or malicious interference.

- **Inequitable   Deployment:** There's a risk that smart infrastructure investments will prioritize wealthy urban areas, leaving rural or underserved communities behind.

Driving Toward a Connected Future

The synergy between autonomous vehicles and smart infrastructure has the potential to redefine mobility, making it safer, more efficient, and more sustainable. By investing in smart cities, integrating autonomous vehicles with public transportation, and leveraging technologies like 5G and IoT, we can create a future where transportation is seamlessly connected to the needs of society.

However, achieving this vision requires more than just technology—it demands a collective effort to rethink how we design cities, manage resources, and prioritize inclusivity. The journey to a connected future is as much about innovation as it is about collaboration, ensuring that the benefits of autonomy are shared by all.

Chapter 9: The Promise of Safer Roads

The open road has always been a place of both freedom and risk. For as long as we've been driving, we've been grappling with the consequences of human error—momentary lapses in judgment that can lead to life-altering accidents. Autonomous vehicles hold the promise of a future where the road is not only a means of connection but also a safer environment for everyone. By replacing human drivers with machines programmed for precision, we stand on the brink of a transformative reduction in traffic accidents and fatalities.

How Autonomous Vehicles Could Reduce Accidents and Save Lives

Every year, millions of people are injured or killed in traffic accidents worldwide. According to the World Health Organization, approximately 1.3 million lives are lost annually due to road traffic crashes. The vast majority of these accidents—over 90%—are attributed to human error. These errors range from distracted driving and speeding to driving under the influence and misjudgment of road conditions.

Autonomous vehicles are designed to eliminate these human failings. Equipped with advanced sensors and algorithms, they maintain constant awareness of their surroundings, react faster than any human could, and adhere strictly to traffic laws.

- **Elimination of Distracted Driving:** Autonomous systems are immune to distractions. They don't text, adjust the radio, or become drowsy. Their focus remains unwavering, 360 degrees around the vehicle, 100% of the time.

- **Consistent Compliance with Traffic Laws:** Self-driving cars are programmed to obey speed limits, stop signs, and traffic signals without exception. This adherence reduces the likelihood of accidents caused by reckless driving behaviors.

- **Advanced Reaction Times:** Machines can process information and respond in milliseconds. In situations where a human might hesitate or react too slowly, an autonomous vehicle can take swift action to prevent a collision.

- **Enhanced Sensing Capabilities:** Autonomous vehicles can utilize technologies like LIDAR, radar, and infrared cameras to see through darkness, fog, or rain—conditions that often impair human drivers.

By addressing the root causes of most accidents,

autonomous vehicles have the potential to make our roads dramatically safer. The ripple effect of this safety extends beyond lives saved—it can reduce medical costs, lower insurance premiums, and alleviate the emotional toll on families and communities.

Data on Human Error vs. Machine Precision

To appreciate the potential impact of autonomous vehicles, it's essential to examine the data comparing human performance to machine capabilities.

- **Reaction Time:** The average human reaction time to visual stimuli is about 1.5 seconds. In contrast, an autonomous vehicle can process and react to data in milliseconds, allowing it to brake or swerve much more quickly than a human driver.

- **Perception Limitations:** Humans have blind spots and limited night vision. Autonomous vehicles use multiple sensors to eliminate blind spots and can see in conditions that would challenge human eyesight.

- **Attention Span:** Studies show that drivers can be distracted up to 50% of the time, whether due to fatigue, phone usage, or other in-car activities. Machines do not suffer from lapses in attention.

- **Error Rates:** While humans can make

unpredictable mistakes due to emotions or misjudgments, machines operate based on precise calculations and pre-defined rules.

However, it's important to note that autonomous vehicles are not infallible. They rely on complex software and hardware systems that can malfunction or encounter situations they are not programmed to handle. Yet, the consistent precision of machines offers a significant advantage over the variability of human behavior.

Challenges in Unpredictable Environments

Despite their advanced capabilities, autonomous vehicles face significant challenges when operating in unpredictable environments.

- **Complex Urban Settings:** Cities present a myriad of unpredictable scenarios—jaywalking pedestrians, erratic cyclists, construction zones, and emergency vehicles. While humans can often navigate these situations using intuition and experience, machines may struggle without explicit programming.

- **Adverse Weather Conditions:** Heavy rain, snow, and fog can interfere with sensor performance. For instance, LIDAR signals can be scattered by raindrops, and cameras may have reduced visibility.

- **Interpretation of Human Behavior:** Subtle cues, like a pedestrian making eye contact with a driver or a traffic officer's hand signals, are difficult for machines to interpret. Understanding intent remains a complex challenge for AI.

- **Infrastructure Variability:** Not all roads are well-maintained or clearly marked. Autonomous vehicles may have difficulty operating in areas with faded lane lines, potholes, or unconventional road designs.

- **Cybersecurity Risks:** Autonomous vehicles are susceptible to hacking and cyberattacks, which could compromise their safety systems.

- **Moral and Ethical Decisions:** As discussed in previous chapters, machines may struggle with ethical dilemmas that require instantaneous moral judgments.

Addressing these challenges requires ongoing research and development. Developers are working on improving sensor technologies, enhancing AI decision-making algorithms, and creating more robust systems that can handle the unpredictability of real-world driving.

A Collaborative Path Forward

The promise of safer roads is a compelling argument for

the adoption of autonomous vehicles, but realizing this promise depends on collaboration among technologists, policymakers, and the public.

- **Continuous Improvement:** Companies must commit to rigorous testing and refinement of their systems, learning from each mile driven and each scenario encountered.

- **Regulatory Support:** Governments need to establish safety standards and certification processes that ensure autonomous vehicles meet stringent safety criteria before widespread deployment.

- **Public Trust:** Building trust with consumers is crucial. Transparency about the capabilities and limitations of autonomous vehicles can help manage expectations and encourage acceptance.

- **Infrastructure Investment:** Upgrading roadways to better accommodate autonomous vehicles—such as implementing smart traffic signals and clear road markings—can enhance their performance and safety.

The Road to a Safer Future

While challenges remain, the potential benefits of autonomous vehicles in reducing accidents and saving lives are too significant to ignore. The shift from human drivers to machine precision represents a monumental step forward in road safety. It offers a future where the tragedy of traffic fatalities could become a rarity rather than a daily occurrence.

As we continue down this path, it's essential to maintain a balanced perspective—celebrating the advancements while diligently working to overcome the obstacles. The journey toward safer roads is not solely about technology; it's about our collective commitment to creating a transportation system that values every life.

Chapter 10: Economic Impacts and Industry Disruption

The advent of autonomous vehicles represents more than just a technological shift—it's an economic revolution poised to reshape industries, create new opportunities, and displace traditional roles. As self-driving technology advances, it is disrupting long-standing business models, forcing companies to adapt, and redefining the way we think about work and mobility.

Job Creation and Displacement: Trucking, Taxis, and Delivery

One of the most immediate and visible economic impacts of autonomous vehicles will be on the job market, particularly in industries reliant on human drivers.

Displacement of Driving Jobs

- **Trucking   Industry:**
  Truck driving is one of the largest occupations in countries like the United States, employing millions of people. Autonomous trucks, capable of long-haul journeys without rest breaks, threaten to replace a significant portion of these jobs.

While drivers may still be needed for short distances and complex maneuvers, the transition could leave many without viable employment.

- **Taxis and Ridesharing:** Companies like Uber and Lyft are heavily investing in autonomous technology, with the ultimate goal of replacing human drivers. Autonomous ridesharing fleets could operate more efficiently and cost-effectively, but they also put millions of gig-economy drivers at risk of job loss.

- **Delivery Services:** Companies like Nuro and Amazon are already deploying autonomous delivery vehicles and drones. These innovations could revolutionize last-mile logistics while simultaneously reducing the demand for delivery drivers.

Creation of New Jobs

While autonomous vehicles will displace some roles, they will also create new opportunities:

- **AI Development and Maintenance:** As self-driving technology evolves, there will be growing demand for engineers, programmers, and technicians to develop, monitor, and repair these systems.

- **Infrastructure   Upgrades:**
  Autonomous vehicles require smarter roadways
  and urban planning, creating jobs in construction,
  engineering, and telecommunications.

- **Fleet   Management:**
  As autonomous vehicles shift from private
  ownership to shared fleets, companies will need
  professionals to manage operations, logistics, and
  customer service.

The net effect on employment will depend on how
quickly industries adapt and whether displaced workers
are retrained for emerging opportunities.

How Traditional Carmakers Are Adapting (or Not)

The rise of autonomous vehicles has forced legacy
carmakers to rethink their strategies. Some have
embraced the shift, while others have struggled to keep
pace with the rapid changes.

Carmakers Leading the Charge

- **General   Motors   (GM):**
  GM's subsidiary, Cruise, has become a
  frontrunner in the race to develop autonomous
  vehicles. By investing heavily in electric,
  autonomous fleets, GM has positioned itself as a

key player in urban mobility.

- **Ford:**
Ford has partnered with Argo AI to develop autonomous technology and is exploring ways to integrate self-driving systems into commercial applications like delivery and ridesharing.

- **Volkswagen and Audi:**
The Volkswagen Group has committed billions to autonomous and electric vehicle development, recognizing that these technologies are the future of transportation.

Carmakers Lagging Behind

Some traditional automakers have been slower to adapt, either due to financial constraints or an over-reliance on existing business models. These companies risk falling behind as competitors and tech companies dominate the autonomous space.

The Shift from Products to Services

For decades, carmakers have focused on selling vehicles to consumers. Autonomous vehicles are driving a shift toward mobility-as-a-service (MaaS), where automakers operate fleets of self-driving cars rather than selling individual units. This transition represents both a challenge and an opportunity for legacy manufacturers.

New Opportunities for Tech Companies and Startups

While traditional carmakers face disruption, the rise of autonomous vehicles has opened doors for tech companies and startups to enter the transportation industry.

Tech Giants

- **Google   (Waymo):**
  Waymo's early investment in autonomous technology has given it a significant lead, allowing it to deploy robotaxi services in select cities.

- **Apple:**
  While Apple's autonomous vehicle project, codenamed "Project Titan," remains shrouded in secrecy, the company's resources and expertise in hardware and software position it as a potential game-changer.

- **Amazon:**
  Amazon is exploring autonomous delivery solutions, leveraging its logistics network and AI capabilities to revolutionize e-commerce transportation.

Startups: The Innovators

Startups have been instrumental in pushing the boundaries of autonomous technology. Companies like

Zoox, Aurora, and Nuro are redefining what autonomous vehicles can be and where they can operate. Their agility and willingness to take risks give them an edge over traditional players.

The Ecosystem of Autonomous Vehicles

The autonomous vehicle revolution has spawned an entire ecosystem of supporting industries:

- **Data Analytics:** Startups specializing in AI training, data labeling, and real-time analytics are thriving.

- **Infrastructure Providers:** Companies focused on smart city infrastructure, 5G connectivity, and vehicle-to-everything (V2X) technology are essential to enabling autonomous vehicles.

- **Cybersecurity Firms:** As autonomous vehicles become more connected, protecting them from cyberattacks has become a critical priority.

A Disrupted Yet Promising Future

The economic impact of autonomous vehicles is a double-edged sword. While they promise efficiency, cost savings, and new industries, they also threaten to disrupt livelihoods and long-standing business models. The key to navigating this transition lies in proactive planning, collaboration between industries, and a commitment to retraining and education.

Governments, companies, and communities must work together to ensure that the benefits of autonomous vehicles are widely shared, minimizing the disruption while maximizing the opportunities. The autonomous revolution is not just about technology—it's about shaping a new economic landscape that balances innovation with inclusion.

Chapter 11: Environmental Implications

As the world faces mounting environmental challenges, the transportation sector—responsible for nearly a quarter of global carbon dioxide emissions—has come under scrutiny. Autonomous vehicles (AVs) promise to revolutionize mobility not only by improving safety and convenience but also by offering potential environmental benefits. Yet, this promise is not without complications. The shift to autonomy brings new challenges related to resource use, energy demands, and the materials that power these cutting-edge technologies.

Potential to Reduce Emissions and Improve Efficiency

One of the most compelling arguments for autonomous vehicles is their potential to reduce greenhouse gas emissions and enhance energy efficiency. Several mechanisms drive this optimism:

- **Efficient Driving Patterns:** Autonomous systems excel at maintaining optimal speeds, accelerating smoothly, and minimizing unnecessary stops—behaviors that maximize fuel efficiency and reduce emissions. Unlike human drivers, AVs avoid inefficient habits like aggressive braking or speeding.

- **Traffic Flow Optimization:** Through vehicle-to-vehicle (V2V) and vehicle-to-infrastructure (V2I) communication, autonomous vehicles can anticipate traffic conditions and adjust routes dynamically. This reduces congestion and idling, which are major contributors to urban air pollution.

- **Shared Mobility and Reduced Car Ownership:** The rise of autonomous fleets may shift society away from private car ownership toward shared mobility models. A fleet of shared AVs could serve multiple passengers in a single trip, significantly reducing the number of vehicles on the road and cutting overall emissions.

- **Integration with Public Transportation:** Autonomous vehicles can complement public transit systems by serving as first-mile/last-mile solutions, making it easier for people to use buses, trains, and subways. This integration reduces dependency on individual cars.

Challenges of Resource Use, Including Rare-Earth Materials

While AVs offer environmental benefits, their production and operation present significant resource challenges, particularly in the context of materials.

- **Rare-Earth   Metals:** Sensors, batteries, and electric motors in autonomous vehicles rely heavily on rare-earth elements like lithium, cobalt, and neodymium. Mining these materials is energy-intensive, environmentally destructive, and often associated with unethical labor practices.

- **Battery   Production:** As most AVs are expected to be electric, their batteries pose unique challenges. Lithium-ion batteries require significant amounts of energy to manufacture and can lead to pollution if not properly recycled.

- **Material   Scarcity:** The demand for critical materials may exceed supply as the industry scales, leading to geopolitical tensions and price volatility. This reliance on finite resources underscores the need for innovations in battery technology and materials recycling.

- **Lifecycle   Emissions:**
  The environmental impact of AVs extends beyond
  their operation. Manufacturing, maintenance, and
  eventual disposal all contribute to the vehicle's
  total carbon footprint. Without sustainable
  practices, these factors could offset the emissions
  reductions achieved during use.

The Interplay Between Autonomy and Electric Vehicles

The convergence of autonomous technology and
electric vehicles (EVs) represents a powerful
opportunity to address environmental concerns. Many
of the advantages of AVs are amplified when paired with
electric powertrains.

- **Energy   Synergy:**
  Electric motors are inherently more efficient than
  internal combustion engines, and their
  compatibility with autonomous systems creates an
  eco-friendly feedback loop. Autonomous EVs can
  optimize charging patterns and energy use,
  further reducing emissions.

- **Smart   Charging   Infrastructure:**
  Autonomous EVs can take advantage of off-peak
  charging times, minimizing the strain on electrical

grids and integrating seamlessly with renewable energy sources. For instance, AVs could charge during periods of high solar or wind energy production, enhancing grid stability and sustainability.

- **Fleet   Electrification:** Shared autonomous fleets are likely to adopt electric powertrains as a cost-effective and environmentally friendly solution. This shift could accelerate the transition to EVs, leading to widespread adoption and greater reductions in emissions.

- **Challenges   in   Battery   Demand:** While the pairing of autonomy and electrification is promising, the demand for EV batteries will surge, exacerbating the resource challenges mentioned earlier. Addressing these challenges requires investments in recycling technologies, sustainable mining practices, and alternative battery chemistries.

Navigating the Environmental Trade-Offs

The environmental promise of autonomous vehicles is significant, but realizing it requires addressing key trade-offs:

- **Policy Interventions:** Governments can incentivize sustainable practices through subsidies for green technologies, regulations on rare-earth mining, and investments in recycling infrastructure.

- **Industry Innovation:** Automakers and tech companies must prioritize sustainable designs, from using recycled materials in manufacturing to developing longer-lasting batteries with minimal environmental impact.

- **Consumer Education:** Encouraging shared mobility and reducing reliance on single-occupancy vehicles can maximize the environmental benefits of AVs. Public awareness campaigns can help align consumer behavior with sustainable goals.

A Greener Horizon

Autonomous vehicles hold the potential to reshape not just how we travel but how we interact with our environment. By reducing emissions, optimizing efficiency, and integrating seamlessly with electric powertrains, AVs could become a cornerstone of a sustainable future. However, this vision depends on navigating the resource challenges and ensuring that the promise of autonomy aligns with broader environmental goals.

As we journey toward this greener horizon, it's clear that autonomy alone won't solve our climate crisis. Instead, it must be part of a larger effort that includes renewable energy, sustainable materials, and a collective commitment to preserving the planet for future generations.

Chapter 12: Cultural Acceptance and Resistance

The adoption of autonomous vehicles is not just a technological or economic challenge—it's a cultural one. Changing the way people perceive driving, trust machines, and embrace new habits requires overcoming deeply ingrained societal norms and fears. While some see self-driving cars as symbols of progress, others view them as threats to autonomy, privacy, or even their identity. Understanding and addressing this cultural landscape is as crucial as refining the technology itself.

Public Perception of Self-Driving Cars: Trust and Fear

The concept of self-driving cars evokes a mix of excitement and apprehension. While many are drawn to the promise of safer roads and greater convenience, others are skeptical about trusting their lives to machines.

- **Trust in Technology:** For autonomous vehicles to gain widespread acceptance, people need to believe in their safety and reliability. However, high-profile incidents involving self-driving cars—such as accidents during testing—have raised doubts. Public confidence hinges on consistent, flawless performance over time.

- **Fear of Losing Control:** Driving is not just a practical activity; for many, it symbolizes freedom and independence. Handing over control to a machine can feel unsettling, especially for those who take pride in their driving skills or feel uncomfortable with technology.

- **Generational Divide:** Younger generations, accustomed to technology-driven lifestyles, tend to be more open to self-driving cars. Older generations, however, may struggle to trust or adapt to this seismic shift in mobility.

- **Privacy Concerns:** Autonomous vehicles collect vast amounts of data about their passengers, raising fears about surveillance and misuse. People worry that their movements could be tracked, their personal habits recorded, or their data exploited for commercial gain.

Societal Shifts in Car Ownership and Driving Habits

Autonomous vehicles are poised to change not only how we drive but also how we think about owning cars. The shift from individual ownership to shared mobility models could redefine transportation on a societal level.

- **The Decline of Private Ownership:** With the rise of autonomous ridesharing fleets, owning a personal car may no longer be necessary—or desirable. Shared AVs can provide on-demand transportation at a fraction of the cost, eliminating expenses like insurance, maintenance, and parking.

- **Driving as a Lost Skill:** If AVs become the norm, driving may eventually become a forgotten skill, much like horseback riding after the advent of automobiles. While this transition may reduce accidents caused by human error, it also raises questions about dependency on technology.

- **Redefining Commutes:** Autonomous vehicles could transform commutes from stressful, time-wasting experiences into productive or relaxing periods. Passengers might use their travel time to work, read, or simply unwind, further blurring the lines between home and office life.

- **Urban and Rural Divide:**
  In urban areas, the adoption of shared AVs could reduce congestion and free up space previously occupied by parking lots. However, rural communities, where car ownership remains essential, may resist these changes due to limited access to shared mobility services.

How Movies, Media, and Cultural Biases Shape Acceptance

Popular culture has a powerful influence on how people perceive autonomous vehicles, often shaping their expectations, fears, and biases.

- **Dystopian Narratives:**
  Science fiction has long depicted autonomous technology as a double-edged sword. Movies like *I, Robot* and *Minority Report* present self-driving cars as marvels of convenience but also caution against over-reliance on machines. Such portrayals can fuel skepticism and fear, particularly about losing control or the potential for misuse.

- **Positive Portrayals:**
  On the flip side, optimistic depictions, like the friendly autonomous cars in Pixar's *Cars* franchise

or the advanced mobility solutions in *Black Panther's Wakanda*, inspire confidence in the potential of this technology.

- **Media   Sensationalism:** News coverage of autonomous vehicles often highlights accidents or controversies, even if such incidents are rare. This selective focus can skew public perception, making AVs seem riskier than they are.

- **Cultural   Biases   and   Norms:** Acceptance of autonomous vehicles varies by culture. In countries with a strong car culture, like the United States or Germany, people may resist giving up the steering wheel. In contrast, societies that prioritize public transportation and urban efficiency, like Japan or Singapore, may embrace AVs more readily.

Bridging the Gap: Building Trust and Encouraging Adoption

To overcome cultural resistance, the autonomous vehicle industry must engage directly with public concerns and aspirations.

- **Education   and   Transparency:** Educating the public about how AVs work, their safety measures, and their limitations can build trust. Transparency about accident investigations, data usage, and system updates fosters confidence in the technology.

- **Demonstrating Benefits:** Pilot programs that showcase the tangible benefits of AVs—such as reduced traffic, lower emissions, and improved accessibility for people with disabilities—can help overcome skepticism.

- **Cultural Adaptation:** Tailoring autonomous vehicle designs and features to fit cultural preferences and norms can increase acceptance. For example, in car-centric cultures, AVs might include manual override options to give drivers a sense of control.

- **Positive Media Campaigns:** Highlighting success stories, such as lives saved or communities improved by AVs, can counterbalance negative narratives. Partnering with trusted influencers or community leaders can also help build credibility.

A Cultural Journey

The road to widespread acceptance of autonomous vehicles is as much about psychology and culture as it is about technology. Overcoming fears, adapting to new norms, and reshaping deeply held values will take time, effort, and empathy. By addressing these cultural challenges head-on, the autonomous vehicle revolution can become not just a technical success but a social one as well.

Chapter 13: Autonomy Around the World

The rise of autonomous vehicles is a global phenomenon, but the path to implementation varies widely depending on regional infrastructure, cultural attitudes, and government policies. While some nations are forging ahead as pioneers, others face significant hurdles. By examining case studies and global trends, we can understand how different regions are tackling the challenges of autonomy and what lessons can be learned from their experiences.

Case Studies: Autonomous Driving in the US, Europe, and Asia

United States: The Testing Ground for Innovation

The U.S. has become a hub for autonomous vehicle testing and development, thanks to its mix of entrepreneurial spirit, technological expertise, and regulatory experimentation. States like California, Arizona, and Texas have become hotbeds for innovation, hosting companies like Waymo, Tesla, and Aurora.

- **Key Features:**

    o Open road testing in diverse environments, from urban centers to highways.

    o Ridesharing pilots, such as Waymo's robotaxi service in Phoenix.

    o A decentralized regulatory approach allowing states to experiment independently.

- **Challenges:**

    o Regulatory inconsistencies across states.

    o Public skepticism fueled by high-profile accidents.

Europe: A Unified Vision with Ethical Oversight

Europe's approach to autonomous driving is defined by its emphasis on safety, ethics, and environmental sustainability. Countries like Germany, Sweden, and the Netherlands lead the charge, supported by EU-wide initiatives like the European New Car Assessment Programme (Euro NCAP).

- **Key Features:**

    o Integration with existing public transit systems, particularly in countries like Sweden.

- Advanced infrastructure, such as smart highways in the Netherlands designed for AVs.

- Strong data protection frameworks, including GDPR, ensuring ethical use of AI.

- **Challenges:**

  - Bureaucratic delays in implementing cross-border regulations.

  - High costs associated with upgrading infrastructure.

Asia: Centralized Vision, Rapid Deployment

Asia is home to some of the fastest advancements in autonomous technology. Countries like China, Japan, and South Korea are leveraging their centralized governance and tech-forward cultures to push the boundaries of self-driving innovation.

- **China:**

  - Testing autonomous fleets in cities like Shenzhen and Beijing.

  - Integration with smart city initiatives to optimize urban mobility.

  - Aggressive government funding for AV startups like Baidu Apollo and Pony.ai.

- **Japan:**

    o Focusing on autonomy for aging populations and rural areas.

    o Introducing Level 3 autonomous vehicles like the Honda Legend.

- **South Korea:**

    o Investing in 5G-enabled autonomous infrastructure.

    o Hyundai's commitment to developing autonomous electric vehicles.

- **Challenges:**

    o Managing rapid urbanization and ensuring equitable access in rural areas.

    o Balancing innovation with privacy concerns.

Infrastructure and Cultural Challenges in Developing Nations

While developed nations are setting the pace, developing countries face unique challenges in adopting autonomous vehicles. These hurdles include:

Infrastructure Gaps

- **Road Quality:** Poorly maintained roads, lack of clear markings, and inconsistent signage hinder the effectiveness of autonomous systems.

- **Smart City Integration:** Autonomous vehicles rely on advanced infrastructure like smart traffic signals and vehicle-to-infrastructure (V2I) communication, which are often lacking in developing regions.

Economic Barriers

- **Cost of Deployment:** The high cost of autonomous vehicles and supporting infrastructure limits accessibility.

- **Reliance on Human Labor:** Many developing nations depend on driving-related jobs, making the transition to autonomy socially and economically challenging.

Cultural and Social Dynamics

- **Trust in Technology:** Skepticism toward new technology can slow adoption.

- **Driving Norms:** In regions with chaotic or informal driving practices, autonomous systems may struggle to adapt.

Despite these obstacles, developing nations have opportunities to leapfrog traditional mobility systems by integrating AVs into future infrastructure planning. Partnerships with global companies and governments can also help bridge the gap.

What We Can Learn from Global Pioneers

The diversity of approaches to autonomous vehicles worldwide provides valuable insights:

- **Collaboration Is Key:** Europe's harmonized regulations demonstrate the value of collaboration among governments, industry, and academia. Shared standards make cross-border adoption more feasible.

- **Infrastructure Investment Pays Off:** The Netherlands and South Korea show how early investment in smart roads and urban planning can accelerate autonomous adoption.

- **Centralized Governance Enables Speed:** China's rapid progress highlights the advantages of centralized decision-making, particularly in funding and scaling pilot programs.

- **Tailoring Solutions to Local Needs:** Japan's focus on autonomy for aging populations and rural areas demonstrates the importance of addressing specific societal challenges.

- **Public Engagement Matters:** U.S. companies are learning that public trust is as important as technological capability. Transparent communication and pilot programs that demonstrate safety and reliability can build acceptance.

A Global Movement, A Shared Future

The path to autonomy is as diverse as the regions pursuing it. While some countries are blazing trails with cutting-edge innovation and infrastructure, others face barriers that require tailored solutions. By learning from global pioneers and fostering international collaboration, the autonomous vehicle revolution can become a shared journey—one that brings safety, efficiency, and opportunity to every corner of the world.

Chapter 14: Hurdles and Setbacks

The road to autonomy has not been without its challenges. Despite remarkable progress, the development and deployment of autonomous vehicles (AVs) have faced significant hurdles. From real-world incidents to delays in technological advancements and the complexities of handling rare, unpredictable situations, these obstacles highlight the immense difficulty of creating vehicles that can operate as safely and reliably as—or better than—humans.

Real-World Incidents Involving Autonomous Vehicles

High-profile incidents involving autonomous vehicles have underscored the risks and limitations of current technologies. While these incidents are relatively rare, they have drawn significant public and regulatory attention.

- **Uber's Fatal Crash (2018):** One of the most infamous incidents occurred in Tempe, Arizona, when an Uber self-driving test vehicle struck and killed a pedestrian. Investigations revealed that the vehicle's sensors detected the pedestrian but failed to classify her as a threat in time to react. Compounding the issue, the human safety driver was distracted at the moment of impact. This tragedy highlighted the dangers of over-reliance on partially autonomous systems.

- **Tesla Autopilot Crashes:** Tesla's Autopilot system, classified as Level 2 autonomy, has been involved in several crashes, often due to drivers misunderstanding its capabilities and failing to remain attentive. These incidents have raised questions about how automakers market their systems and whether more stringent oversight is needed.

- **Waymo's Minor Collisions:** Even industry leader Waymo has faced challenges, with reports of its vehicles being involved in minor accidents. While most were caused by human drivers in other cars, these incidents illustrate the difficulty of integrating AVs into a mixed-traffic environment.

Public Perception Impact

Each incident damages public trust and fuels skepticism, even if autonomous vehicles statistically outperform human drivers. Clear communication about the capabilities and limitations of these systems is crucial to rebuilding confidence.

Delays in Technological Development and Deployment

The promise of fully autonomous vehicles has been met with delays, as developers encounter unexpected technical and logistical challenges.

- **Unrealistic Timelines:** Early predictions of widespread autonomous adoption by the early 2020s have proven overly optimistic. Elon Musk, for example, repeatedly projected that Tesla would achieve full self-driving capabilities, but those timelines have consistently been pushed back.

- **Complexities of Urban Environments:** Cities present a maze of challenges, from unpredictable pedestrians to chaotic traffic patterns. Developers have found that creating systems capable of navigating these environments safely is far more difficult than anticipated.

- **Regulatory Uncertainty:** Governments have struggled to keep pace with technological advancements, creating a patchwork of regulations that slow progress.

Companies often face lengthy approval processes for testing and deployment.

- **Costs and Scalability:** Developing, testing, and deploying autonomous systems is prohibitively expensive. The costs of LIDAR sensors, high-definition mapping, and extensive testing fleets make it difficult for smaller companies to compete and delay the scaling of AVs.

The Challenge of Redundancy and Edge Cases

Redundancy: Building Fail-Safe Systems

For autonomous vehicles to be truly reliable, they must have redundancy built into every critical system. This means that if one component fails, another can immediately take over to prevent accidents.

- **Hardware Redundancy:** Multiple sensors (e.g., LIDAR, radar, and cameras) must overlap in functionality to ensure the vehicle can "see" its surroundings even if one system malfunctions.

- **Software Redundancy:** Backup algorithms are required to handle unexpected scenarios or errors in primary decision-making systems. Developing these layers of fail-safe mechanisms is both complex and resource-intensive.

Edge Cases: The Unpredictable and Unprogrammed

Edge cases—rare or unusual scenarios that fall outside typical driving conditions—pose one of the greatest challenges for AVs.

- **Examples of Edge Cases:**

    o A child chasing a ball into the street.

    o Construction zones with ambiguous lane markings.

    o A driverless vehicle encountering a horse-drawn carriage or other uncommon vehicles.

    o Unusual weather phenomena like sudden sandstorms or hail.

- **The Long Tail Problem:** While AVs can be trained to handle the most common driving scenarios, programming them to account for every possible edge case is practically impossible. Developers rely on machine learning and simulation to prepare for these situations, but real-world deployment reveals gaps that require further refinement.

A Road Still Under Construction

The hurdles faced by autonomous vehicles are not insurmountable, but they remind us of the complexity of the task at hand. Achieving full autonomy requires not only technological innovation but also patience, rigorous testing, and collaboration across industries.

- **Learning from Incidents:** Each accident or setback provides valuable data, allowing developers to refine their systems and improve safety.

- **Evolving Timelines:** While early predictions proved overly ambitious, the industry has adopted a more measured approach, focusing on incremental progress and realistic goals.

- **Building Trust Through Transparency:** Public and regulatory trust is paramount. Companies must be transparent about their challenges, communicate their limitations, and demonstrate accountability.

Turning Setbacks into Opportunities

Hurdles and setbacks are inevitable in any technological revolution. They reflect the scale of ambition driving the autonomous vehicle industry. Each challenge overcome brings us closer to a future where

AVs not only match but exceed human driving capabilities, making our roads safer and more efficient for all.

Chapter 15: The Future of Autonomy

The road to autonomy is far from complete, but the pace of progress suggests that the next decade will bring transformative changes to transportation. With advancements in artificial intelligence (AI), infrastructure, and policy, the vision of a fully autonomous world is becoming clearer. This chapter explores the likely trajectory of self-driving innovation, the broader implications of AI on mobility, and a glimpse into what a future shaped by autonomy might look like.

Predictions for the Next Decade of Self-Driving Innovation

While fully autonomous vehicles (Level 5) are not expected to dominate the roads immediately, significant advancements are likely in the coming decade.

Short-Term Developments (1–3 Years)

- **Expanded Testing and Deployment:** Companies like Waymo, Tesla, and Cruise will continue refining their autonomous systems, with more pilot programs in urban areas and specific applications like delivery or freight.

- **Wider Availability of Level 3 Vehicles:** Automakers such as Mercedes-Benz and Honda are introducing Level 3 autonomy in select markets, allowing drivers to disengage under certain conditions like highway driving.

- **Increased Regulation:** Governments will implement clearer guidelines and standards for autonomous vehicle testing and deployment, reducing uncertainty for companies and consumers.

- **Focus on Safety and Public Trust:** Public education campaigns and transparent communication will be critical as companies work to rebuild trust after high-profile incidents.

Mid-Term Developments (4–7 Years)

- **Commercialization of Autonomous Fleets:** Ridesharing and delivery services will increasingly rely on autonomous vehicles, reducing costs and increasing efficiency.

- **Integration with Smart Cities:** Autonomous vehicles will become part of larger smart city ecosystems, communicating with infrastructure like traffic lights, parking systems, and public transit hubs.

- **Advancements in Edge Case Handling:**
  AI systems will become better at managing rare and unpredictable scenarios, thanks to continued machine learning and data collection.

Long-Term Developments (8–10 Years)

- **Emergence of Level 4 Autonomy in Everyday Use:**
  Fully autonomous vehicles in controlled environments—like specific urban zones or highways—will become commonplace.

- **Cost   Reductions:**
  Advances in technology and scaling of production will make autonomous vehicles more affordable, driving mass adoption.

- **Policy   Shifts:**
  Governments may begin incentivizing autonomous vehicle use to reduce traffic accidents and emissions, while addressing challenges like job displacement.

How AI Might Redefine Transportation Entirely

AI is the engine driving the autonomous revolution, and its influence on transportation goes beyond self-driving cars. The integration of AI will redefine how we think about mobility and its role in our lives.

- **Personalized Transportation:** AI-driven systems will optimize routes and modes of transportation based on individual preferences, creating a seamless and efficient travel experience.

- **Decentralized Mobility Networks:** Autonomous vehicles will form interconnected fleets, sharing data in real time to optimize traffic flow, reduce congestion, and improve energy efficiency.

- **Accessibility and Inclusivity:** Self-driving technology will enable mobility for those previously excluded, such as the elderly, people with disabilities, and those in rural areas.

- **Sustainability Through AI:** AI will enable dynamic carpooling, fleet management, and energy optimization, reducing the environmental footprint of transportation.

- **Transformation of Logistics:** Autonomous trucks and drones will revolutionize supply chains, making deliveries faster, cheaper, and less resource-intensive.

A Vision of a Fully Autonomous World

Imagine a city where self-driving cars glide silently through the streets, orchestrated by AI systems that optimize every movement. Traffic jams are a thing of the past, parking lots have been repurposed into parks and housing, and accidents are so rare they become an anomaly. In this vision of a fully autonomous world, transportation is no longer a source of stress but an invisible, efficient part of daily life.

Urban Transformations

- Streets are designed for people rather than cars, with expanded sidewalks, bike lanes, and green spaces.

- Smart intersections manage traffic flow seamlessly, reducing wait times and energy consumption.

- Public transit systems are integrated with autonomous shuttles, creating a multimodal network that serves everyone.

Rural Connectivity

- Autonomous vehicles bring mobility to underserved rural areas, connecting communities to jobs, healthcare, and education.

- Freight systems use autonomous trucks to deliver

goods efficiently to remote locations.

Cultural Shifts

- The concept of "driving" fades into history, as younger generations grow up without ever learning to operate a car.

- Owning a car becomes a niche interest, replaced by shared mobility services that prioritize convenience and sustainability.

Global Impact

- A fully autonomous world could save millions of lives by eliminating traffic accidents caused by human error.

- It could dramatically reduce emissions, helping combat climate change and create cleaner, healthier cities.

- Economies will shift, with new industries rising around AI, data management, and fleet services.

Navigating the Road Ahead

The future of autonomy is both exciting and uncertain. While the potential benefits are enormous, the journey will require addressing significant challenges, including ethical dilemmas, regulatory frameworks, and public trust. Collaboration between governments, companies, and communities will be essential to ensure that this vision is realized in a way that benefits everyone.

As we stand on the cusp of this transformative era, one thing is clear: the future of transportation is autonomous, and the road ahead promises to be a journey unlike any other.

Chapter 16: Conclusion: A World in Motion

As the journey through the realm of autonomous vehicles draws to a close, one truth stands out: self-driving technology is not just a technological revolution—it is a societal transformation. The road to autonomy has been long and winding, marked by breathtaking innovation and formidable challenges. Yet, at every turn, it has revealed the potential to reshape how we live, work, and connect.

Reflecting on the Journey of Self-Driving Technology

The story of autonomous vehicles is a testament to human ingenuity. What began as a futuristic vision has become an unfolding reality, driven by advances in artificial intelligence, engineering, and data science. Along the way, we have encountered defining moments—milestones that demonstrate both the progress and the complexity of this endeavor.

From the early prototypes that stumbled in the DARPA challenges to today's fleets of semi-autonomous cars navigating city streets, self-driving technology has evolved in leaps and bounds. Yet, this journey is far from over. Each breakthrough brings new questions, requiring not only technical solutions but also ethical frameworks, regulatory innovation, and cultural shifts.

As we reflect on this journey, it's clear that autonomous vehicles are more than machines; they are symbols of our aspirations for a safer, more efficient, and more connected world.

The Balance of Promise and Peril

Like all transformative technologies, self-driving cars come with both immense promise and significant peril. Striking a balance between these forces is critical to shaping a future where autonomy serves the greater good.

- **The   Promise:** Autonomous vehicles offer the potential to save millions of lives by eliminating traffic accidents caused by human error. They can reduce emissions, optimize urban spaces, and make transportation more inclusive and accessible. For industries, they promise efficiency and innovation, transforming logistics, ridesharing, and public transit.

- **The   Peril:** However, these benefits come with risks. Ethical dilemmas, such as decision-making in life-or-death scenarios, challenge our moral frameworks. Job displacement in sectors like trucking and delivery threatens livelihoods, while concerns over privacy and cybersecurity loom large. Without equitable deployment, autonomous technology could deepen societal divides, leaving rural or underserved communities behind.

Balancing these forces requires foresight, collaboration, and a commitment to aligning technology with humanity's broader values.

How Society Must Adapt to a Changing Way of Life

The shift to autonomy will demand not only technological progress but also societal adaptation. To fully embrace the potential of self-driving vehicles, we must rethink longstanding norms and systems.

Policy and Regulation

Governments must create forward-thinking regulations that address safety, liability, and data privacy. These policies should foster innovation while protecting public interests, ensuring that the benefits of autonomy are distributed equitably.

Education and Workforce Development

Society must prepare for the disruptions caused by automation. Programs to retrain workers in industries affected by self-driving technology are essential. At the same time, education systems should emphasize skills in AI, robotics, and ethics, equipping the next generation to thrive in an autonomous world.

Cultural Shifts

Accepting self-driving technology means redefining our relationship with driving itself. For some, this will involve relinquishing a cherished sense of control; for others, it will mean embracing a new level of trust in machines. Building public confidence will require transparency, consistent performance, and clear communication from developers and policymakers.

Urban and Rural Planning

The integration of autonomous vehicles will reshape our cities and towns. Urban planners must adapt to a world with fewer parking lots, smarter traffic systems, and more efficient use of space. Rural areas, on the other hand, must be included in the deployment of this technology to ensure equal access to its benefits.

A World in Motion

Autonomous vehicles represent a world in motion—both literally and figuratively. They challenge us to rethink what it means to travel, to connect, and to innovate. They force us to confront difficult questions about ethics, equity, and control, while offering solutions to some of the world's most pressing problems.

As we move forward, the choices we make today will define the role of self-driving technology in our future. Will it be a tool for inclusivity and sustainability, or will it deepen divides and create new challenges? The answer depends on our collective ability to navigate this journey with wisdom, empathy, and purpose.

The road ahead is full of promise, but it requires careful navigation. By embracing the possibilities and addressing the challenges with equal determination, we can create a world where autonomy enhances—not replaces—our humanity. In this vision, self-driving technology is not the destination but a vehicle for progress, carrying us toward a safer, smarter, and more connected future.

About the Author

Etienne Psaila, an accomplished author with over two decades of experience, has mastered the art of weaving words across various genres. His journey in the literary world has been marked by a diverse array of publications, demonstrating not only his versatility but also his deep understanding of different thematic landscapes. However, it's in the realm of automotive literature that Etienne truly combines his passions, seamlessly blending his enthusiasm for cars with his innate storytelling abilities.

Specializing in automotive and motorcycle books, Etienne brings to life the world of automobiles through his eloquent prose and an array of stunning, high-quality color photographs. His works are a tribute to the industry, capturing its evolution, technological advancements, and the sheer beauty of vehicles in a manner that is both informative and visually captivating.

A proud alumnus of the University of Malta, Etienne's academic background lays a solid foundation for his meticulous research and factual accuracy. His education has not only enriched his writing but has also fueled his career as a dedicated teacher. In the classroom, just as in his writing, Etienne strives to inspire, inform, and ignite a passion for learning.

As a teacher, Etienne harnesses his experience in writing to engage and educate, bringing the same level of dedication and excellence to his students as he does to his readers. His dual role as an educator and author makes him uniquely positioned to understand and convey complex concepts with clarity and ease, whether in the classroom or through the pages of his books.

Through his literary works, Etienne Psaila continues to leave an indelible mark on the world of automotive literature, captivating car enthusiasts and readers alike with his insightful perspectives and compelling narratives.

Visit www.etiennepsaila.com for more.

www.ingramcontent.com/pod-product-compliance
Lightning Source LLC
LaVergne TN
LVHW021459170326
834004LV00003B/347